W9-CTX-272

STRUCTURES AND MATERIALS

Barbara Taylor

Photographs by Peter Millard

FRANKLIN WATTS
New York • London • Toronto • Sydney

Franklin Watts, Inc.
387 Park Avenue South
New York, NY 10016

Design: Janet Watson

Science consultant: Dr. Bryson Gore

Primary science adviser: Lillian Wright

Series editor: Debbie Fox

Editor: Roslin Mair

Illustrations: Linda Costello

The author and publisher would like to thank the following children for their participation in the photography of this book: Joanna Archer, Colleen Delaney, Karim Dlimi, Charmaine Gentle, Bonita KC, Roshan Meghani, Nishma Patel, Amanda Shannon, Paul Stocker, Kirstie Wallace, Leon Young.

Thanks to Carol Olivier of Kenmont Primary School, Ganga Budhathoki and Mrs. Willan.

Library of Congress Cataloging-in-Publication Data

Taylor, Barbara, 1954-
Structures and materials/ Barbara Taylor.
p. cm. – (Science starters)
Includes index.
Summary: Examines the properties of various building materials and shows through text and specific do-it-youself projects how the structure of roofs, foundations, and bridges relate to that of the human body.
ISBN 0 531-14186-1
1. Strength of materials – Juvenile literature.
[1. Strength of materials.] I. Title. II. Series.
TA405.T288 1991
620.1'12–dc20
91-2282 CIP AC

Printed in Belgium

CONTENTS

This book is all about the shape and strength of buildings and other structures, and the properties of natural and artificial building materials. It is divided into five sections. Each has a different colored triangle at the corner of the page. Use these triangles to help you find the different sections.

These red triangles at the corner of the tinted panels show you where a step-by-step investigation starts.

USING STRUCTURES

How many different buildings can you see in this city?

We use buildings for many purposes such as working, worshipping, storing things, or making electricity. How many other uses can you think of? Apart from buildings, our world is full of other structures. Some, such as animals or rocks, are natural structures. Others, such as bridges or cranes, are made by people.

A structure is put together in a particular way and made of certain materials which are suited to the job it has to do. Structures are not always large, complicated things such as skyscrapers or castles. Shoe boxes and mugs are structures too. They keep their shape when they are used correctly. Boxes are strongest at the four corners. They are useful for storing things because they can be stacked without collapsing.

Another simple, everyday structure is a step ladder. This is an example of a frame, which is a structure made to enclose, border or support something. Other frames include eyeglass frames, stairs, climbing frames and window frames. This step ladder forms a triangular shape which is strong because it is rigid and keeps its shape well.

Have you ever put up a tent? Tents are useful homes for people who travel from place to place or go camping. They also provide instant temporary housing in emergencies. They are lightweight, quick to assemble and easy to pack and carry. Mountaineers use tents made from special artificial materials which are designed to be very light, strong and waterproof. The frame on this tent holds the outer skin away from the inner one, helping to keep the tent dry and warm inside.

Tall towers called pylons support electrical wires high above the ground where the powerful and dangerous electricity is safely out of reach. The crisscross shape of the framework gives extra strength without making it too heavy. Did you know that your bones have a meshwork inside to make them strong but light?

Some structures, such as bridges, are designed to help us cross natural barriers such as rivers or to carry traffic and railroads above busy city streets. Bridges may be made of rope, wood, stone, metal or cement. They have to be able to support heavy weights. The arch of this bridge is a strong structure. You can find out why on page 11.

BUILDING MATERIALS

Years ago, people had very few materials. They made use of natural materials, such as wood and stone, which they found nearby. Today, a great variety of raw materials are in use, and they can be transported more easily. Far more "artificial" materials are made, for example, steel, glass, plastics and cement.

Architects and builders have a wide range of materials to choose from. The type of material they use depends not only on how suitable it is for its function, but also on things such as cost, local environmental conditions and building regulations.

Animals build structures for many reasons such as to escape from predators and to protect their young. Some animal builders, like beavers or birds, use materials in their environment such as twigs and grass. Other animals, such as spiders and bees, produce their own building materials inside their bodies.

This spider is spinning a silken web to catch food. Can you see the silk coming out of the back of its body? Silk is extremely strong. It is also very elastic so it can absorb the impact of a flying insect.

Materials are made up of tiny particles called molecules. Each material has its own kinds of molecules, which fit together in a particular pattern. This gives the material its own unique properties of strength, weight, flexibility, texture and so on.

In terms of structures, strength is very important. Try comparing the different strengths of paper, plastic and aluminum foil by hanging weights from a small strip of each material. Which is the strongest? How does each tear? Do different shapes and amounts of the material affect the results? A force like this which pulls the ends of something apart is called a tension force.

Can you see the metal rods sticking out of the concrete in the picture? The metal rods will snap less easily than the concrete, giving it extra strength, which is why it is called reinforced concrete. Long, thin strips of material, such as rods, ropes and cables, are good at resisting tension forces.

Sometimes, materials have to withstand compression forces, that push or squeeze them from the outside. Flexible, or friable, materials,are easily squeezed out of shape. Hard, brittle materials, such as stone, are good at surviving these forces.

The stone in the center of an arch is a force pushing downward. It is held up by the upward pushing force of stones on either side. An arch has to be firmly supported at the ends to keep it from being squeezed outward by forces passing down through the stones to the ground.

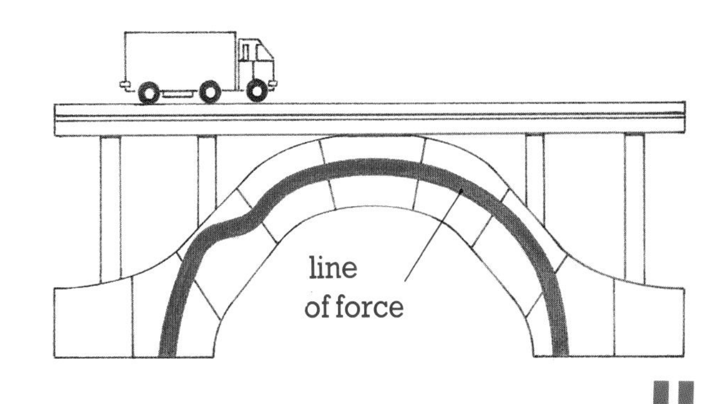

Another important property of materials is weight. Collect a variety of cans and compare their weights. In the picture, a pile of aluminum cans weighs the same as a few steel cans. Aluminum is light but very strong. The material is less dense than steel, good for making into cans.

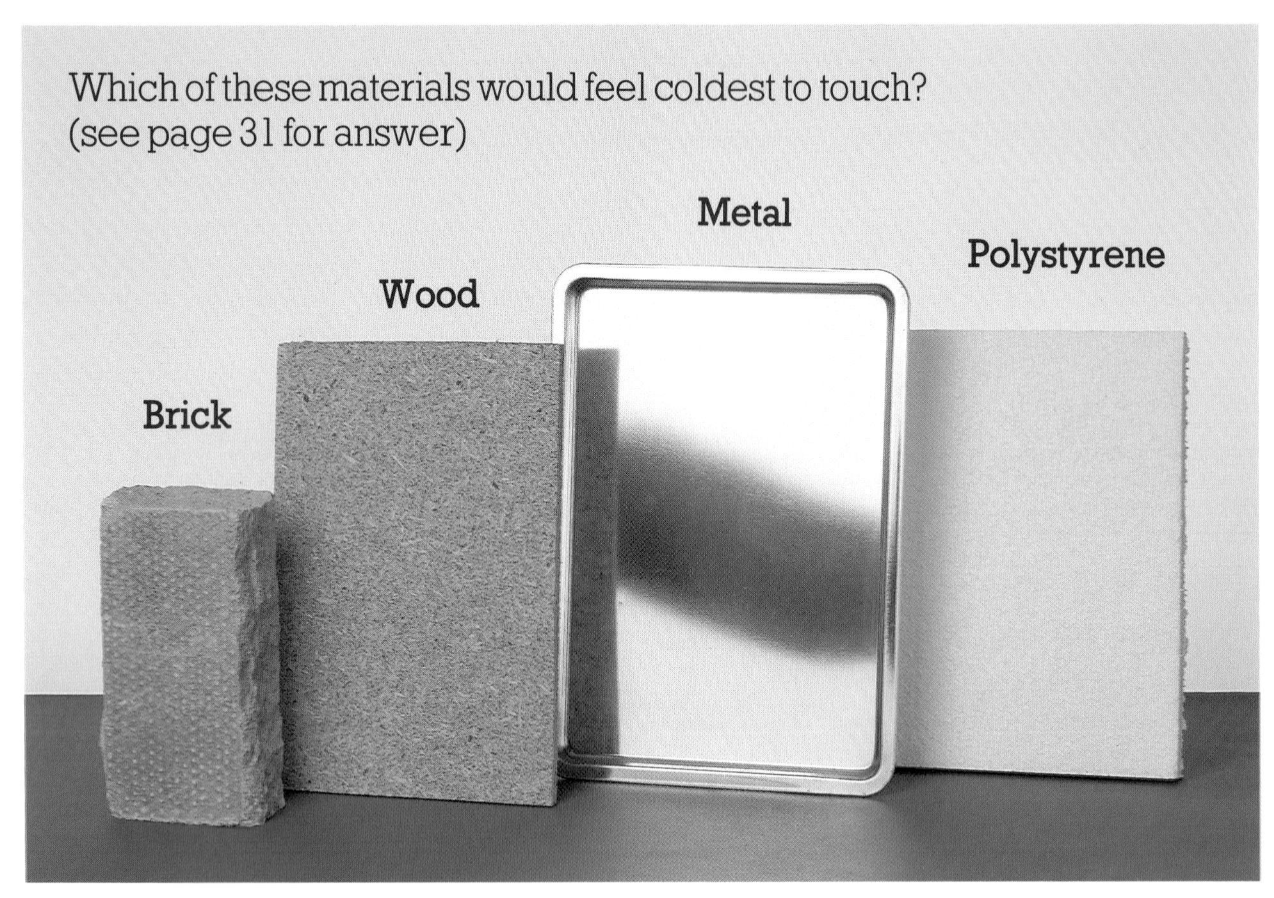

Materials that feel cold carry heat away from your fingers easily. They are said to be good conductors. Materials that feel warm do not carry heat so easily. They are said to be good insulators. Materials that are good insulators are very useful for keeping buildings warm.

The choice of materials for a structure depends to some extent on how long the structure has to last. Some materials, such as wood, can be eaten away by living things, such as bacteria and fungi. They are said to be biodegradable, which means "able to be broken down by living things." Wood can be protected by varnish or paint, which seals the surface and helps to keep it from rotting.

Poisonous gases from power stations and vehicle exhausts mix with water in the air to make rainwater more acidic. This "acid rain" eats away at building stone, gradually making it crumble away. Have any of the buildings in your area been damaged by acid rain?

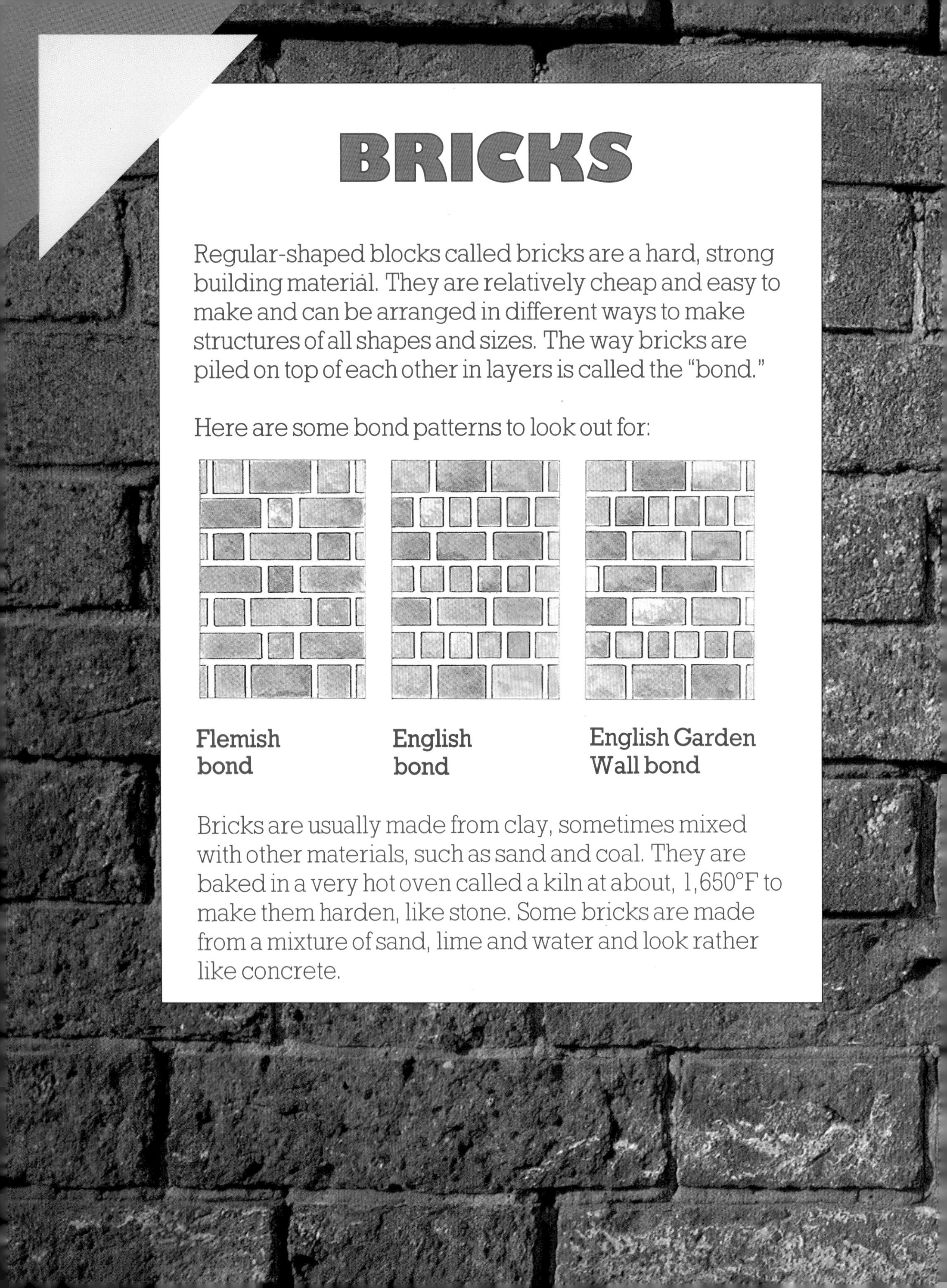

BRICKS

Regular-shaped blocks called bricks are a hard, strong building material. They are relatively cheap and easy to make and can be arranged in different ways to make structures of all shapes and sizes. The way bricks are piled on top of each other in layers is called the "bond."

Here are some bond patterns to look out for:

Flemish bond

English bond

English Garden Wall bond

Bricks are usually made from clay, sometimes mixed with other materials, such as sand and coal. They are baked in a very hot oven called a kiln at about, 1,650°F to make them harden, like stone. Some bricks are made from a mixture of sand, lime and water and look rather like concrete.

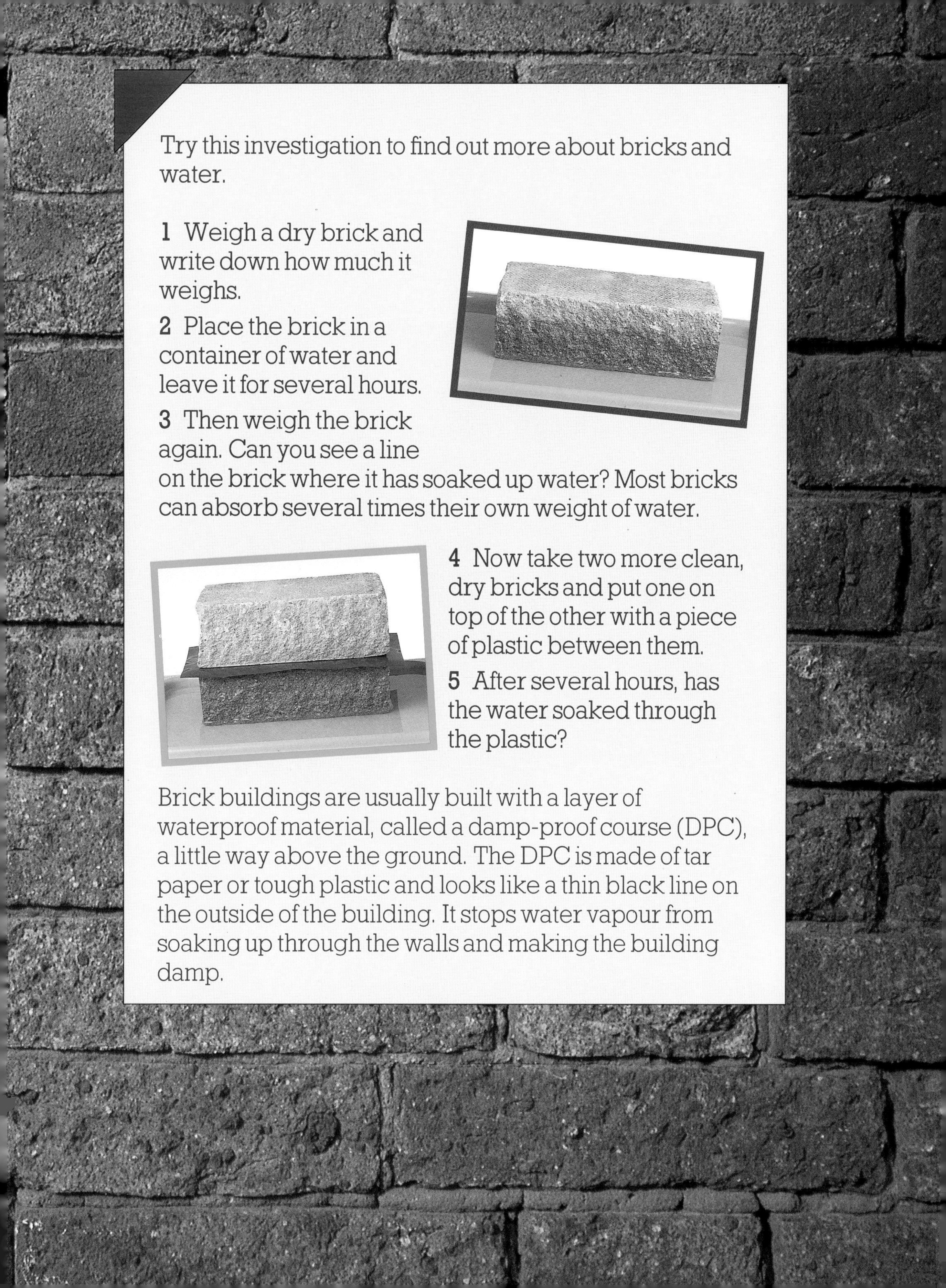

Try this investigation to find out more about bricks and water.

1 Weigh a dry brick and write down how much it weighs.

2 Place the brick in a container of water and leave it for several hours.

3 Then weigh the brick again. Can you see a line on the brick where it has soaked up water? Most bricks can absorb several times their own weight of water.

4 Now take two more clean, dry bricks and put one on top of the other with a piece of plastic between them.

5 After several hours, has the water soaked through the plastic?

Brick buildings are usually built with a layer of waterproof material, called a damp-proof course (DPC), a little way above the ground. The DPC is made of tar paper or tough plastic and looks like a thin black line on the outside of the building. It stops water vapour from soaking up through the walls and making the building damp.

JOINING MATERIALS

How much glue do you need to join two pieces of wood so they will take the weight of two bags of marbles? What happens if the joint gets wet?

Have a look around your home or school and see how many different ways you can find of joining things together. The type of joint depends partly on the material itself. Pieces of wood can sometimes be joined by careful cutting so one piece fits into another.

Metal cannot be joined in this way. If metal pieces need to be taken apart, they can be joined with nuts and bolts. More permanent metal joints are made by melting pieces of metal together at high temperatures. This is called welding.

Bricks are stuck together with a mixture of sand, cement and water, which is called mortar. The wall is stronger if the bricks overlap each other. In this way the joints, which are the weakest part of the wall, are not aligned one on top of the other.

It takes great skill to build a wall without using mortar in the joints. This "dry-stone" wall was built by the Incas in Peru. The stones fit so closely that it is hard to put the blade of a knife between them. The stones were shaped with stone hammers and tested over and over again until the fit was perfect. These ancient walls are heavy enough to stand firm in an earthquake, yet flexible enough to withstand the shock of the ground moving.

WEAK AND STRONG SHAPES

Can you squash a cardboard tube by pressing down on the end? A tube is a very strong shape as long as the pressing force runs along its sides. It is easy to make it buckle if you press the sides inward.

Tubes are good for making structures because they are strong but light. They are used in scaffolding, aluminum walkers, beverage cans, towel bars and swings. The roofs of indoor swimming pools usually have to cover a large area but can only be supported at the edges. A framework of tubes is a good way of supporting a roof like this.

Can you build a tower from drinking straws? How high can you make the tower? What is the best way of arranging the straws to make a firm shape? There are lots of different ways of joining the straws together. Glue or tape makes rigid joints but pipe cleaners or modeling clay can make movable joints. If you pin the straws together, you will be able to turn the straws in different directions and make a flexible structure.

ROOFS

Have a look at the roofs in your area. How many different designs can you find? What covering materials are used? Most small buildings have a sloping roof which is easy to build and allows rain to run off.

Try this investigation to find out more about the forces underneath a roof.

1 Place two large cereal boxes on a table a small distance apart. These represent the walls of a building.

2 Make a roof by opening a book in the middle. See if you can balance it on top of the walls. The downward pushing force of the roof will probably make the walls fall over.

3 Now place two pieces of wood across the walls before you put the roof on. These cross beams spread out the pushing forces from the roof so they are not just acting at one point and the walls do not fall over.

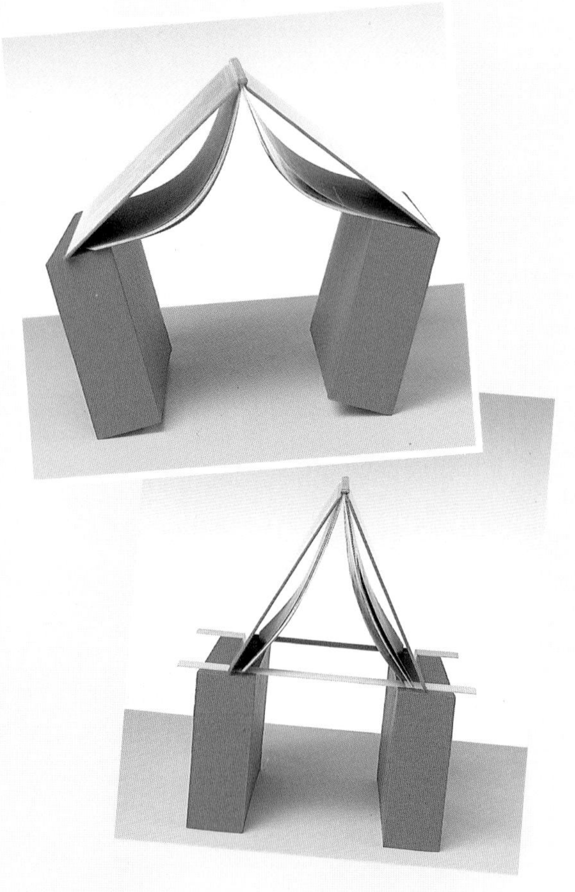

On older buildings, such as cathedrals or castles in Europe, supporting walls called buttresses can be found. These were used to counterbalance the outward pushing forces and prop up the walls. On towers, buttresses are usually at the corners of the walls.

Architects often use dome shapes in large buildings because domes can span large spaces. A dome is good at carrying its own weight. The dome shape spreads out the load so the pushing force of the roof is evenly distributed.

To figure out the strongest shape for a dome, hold a chain up in front of you. The curve gives the outline of a dome, upside down. The force of gravity is dispersed evenly through the length of the chain.

FOUNDATIONS

Have you ever stood on a beach and felt yourself sinking into the sand? All big structures need a firm base or foundation to support their weight and keep them from sinking into the ground, or from being blown over by strong winds. The best foundations are solid rock. Some of the tallest buildings in the world have been built in New York where there is solid rock near the surface. Usually, the rock is too far below the surface to help provide a foundation. Instead, large slabs, or mats of concrete, or pillars of concrete or steel, are used for support. They spread the weight of the building over a bigger area.

Tall skyscrapers are built with some "give" in their structure so they can sway a little in high winds and so are able to move if the ground shakes during an earthquake. A flexible structure is more likely to absorb the shock of an earthquake and avoid collapse. Before a skyscraper is built, scientists build a small-scale model and test it to see how well it stands up to winds and earthquakes.

Try this investigation to find out more about foundations.

1 Half fill a bowl, tray or other container with sand.

2 Press a small coin into the sand. How far does it sink?

3 Now cut out a disk of stiff cardboard and put it under the coin. This spreads out the pushing force, or pressure, of the coin over a larger area so it does not sink in as far.

4 Balance a coin on a pencil that reaches the bottom of the container. When you press on the coin, the pencil transfers the force down to the solid base of the container so the coin cannot sink down.

BRIDGES

The first bridges were natural arches of rock or tree trunks laid across rivers. Today, bridges are built from all kinds of materials and can be made in all shapes and sizes. Like other structures, bridges need firm foundations or they will collapse.

There are many types of bridge. The Golden Gate bridge at the entrance to San Francisco Bay is a suspension bridge in which the deck hangs, or is suspended, from thick steel cables. The cables are held up by tall towers. They are strong enough to carry the weight of the bridge so there is no need for columns or arches underneath. The cables transfer the weight to the towers, and then to the ground. Steel stands up well to tension or stretching forces. Suspension bridges can span very wide waterways.

Try making some different bridges.

1 A beam bridge consists of a horizontal beam which is supported on the ground or by piers at either end. How much weight will your beam bridge support without bending in the middle? To span a long gap, a beam bridge needs many columns or piers for support.

2 A cantilever bridge is usually made up of two sections, each of which has its main support at one end. There is no support in the middle. Sometimes cantilever bridges have a third section in the middle, to make the bridge longer. Cantilever bridges can cover greater distances than simple beam bridges. You can often see cantilever bridges on roadways.

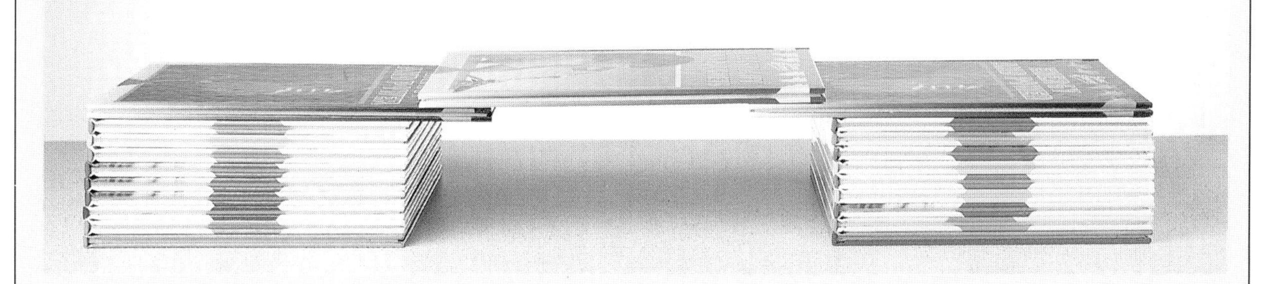

3 Arch bridges carry the weight outward along two curving paths down to the ground. The ground resists the outward pushing force and holds the bridge up. A level roadway may be built on top of a series of arches.

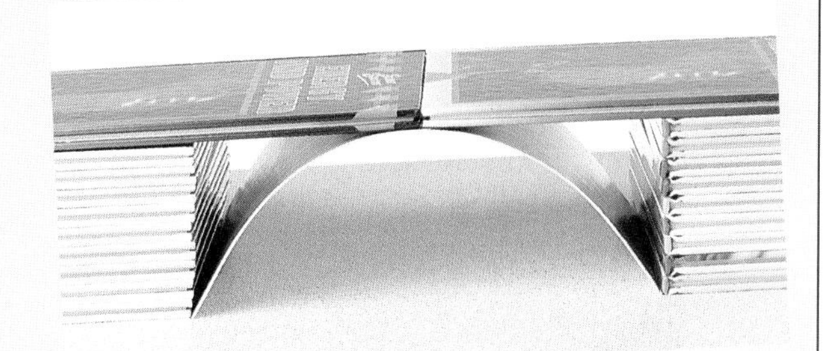

Can you make a long bridge which includes two or more types of bridge?

OURSELVES

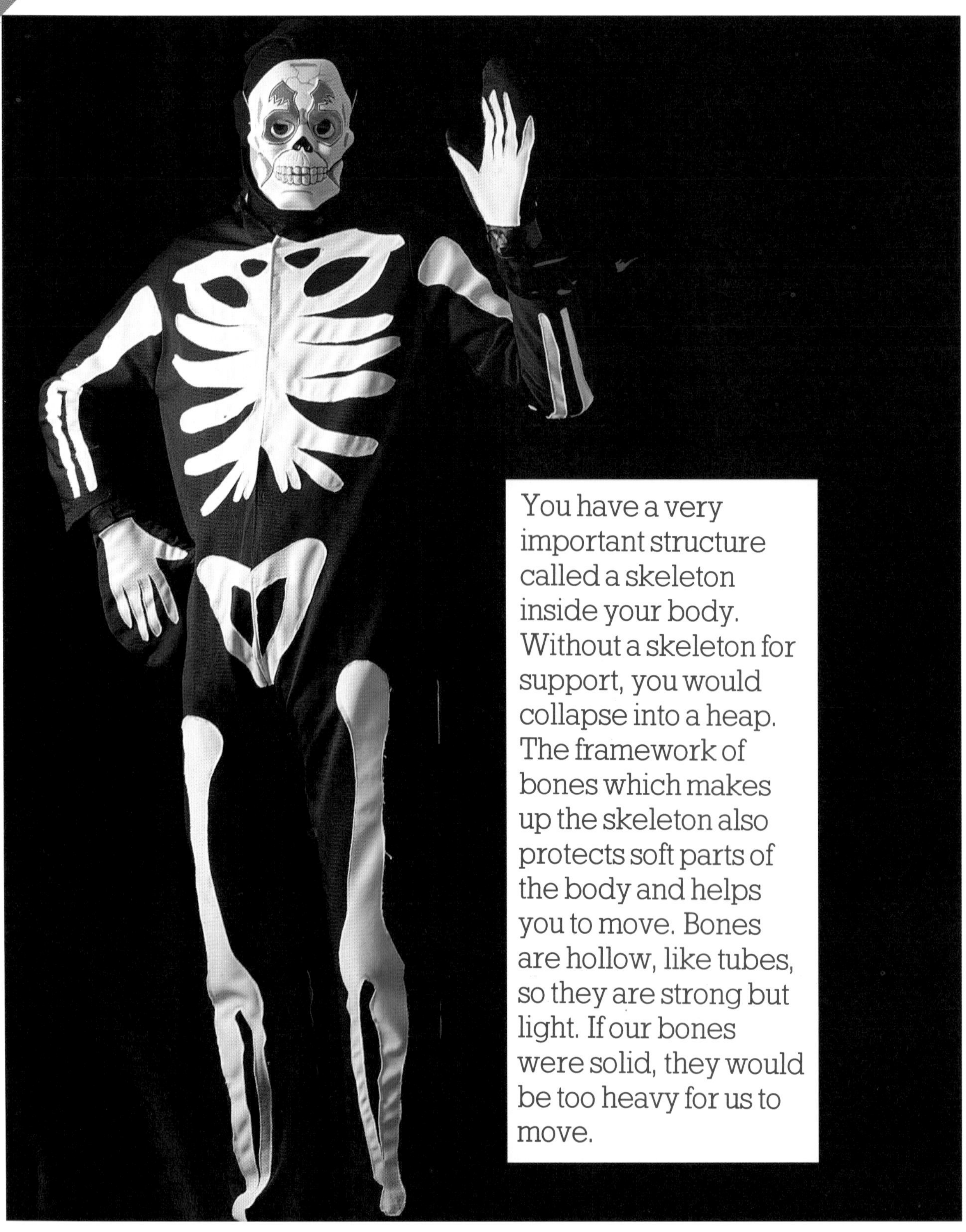

You have a very important structure called a skeleton inside your body. Without a skeleton for support, you would collapse into a heap. The framework of bones which makes up the skeleton also protects soft parts of the body and helps you to move. Bones are hollow, like tubes, so they are strong but light. If our bones were solid, they would be too heavy for us to move.

The twelve pairs of ribs in the chest are tough and springy. They make a strong, cagelike framework to protect the heart and lungs. They also help us to breathe. Muscles pull on the ribs and make the rib-cage expand and contract. This allows the lungs to fill with air and then helps to push air out of the lungs as you breathe out.

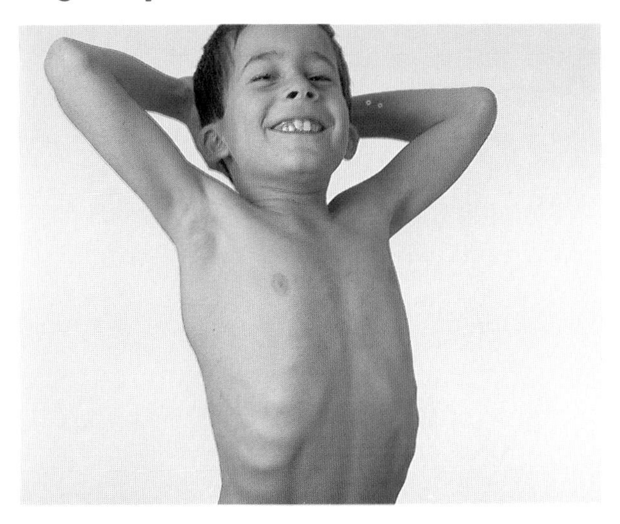

Bones cannot bend, so wherever two bones meet there is a structure called a joint. Joints allow you to move parts of your body. At your elbow there is a hinge joint, which allows you to bend your arm up and down.

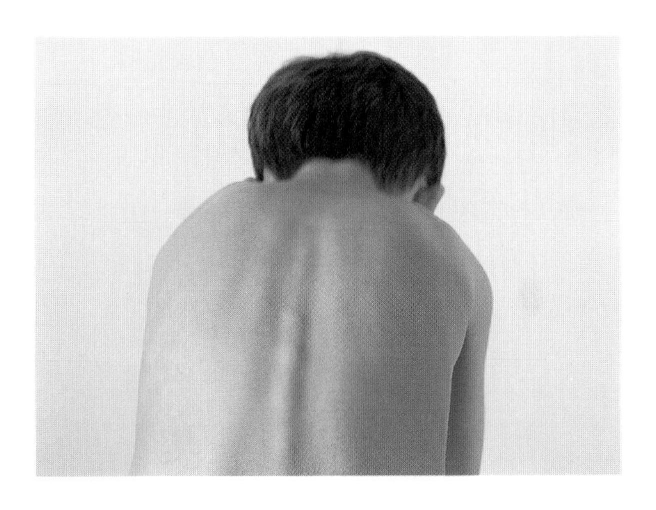

The hollow, dome-shaped skull forms a strong, protective box for the brain, eyes and ears. The main part of the skull, the cranium at the top, is made of eight separate bones.

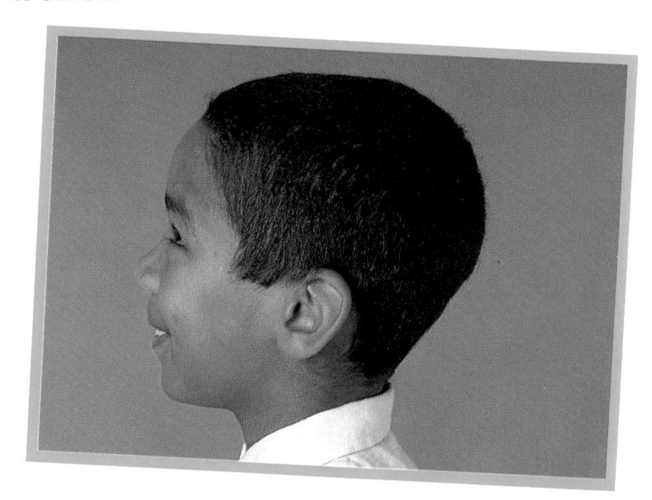

The backbone is a strong support for your whole body. It is made up of many bones called vertebrae. These form a long, bony tube, which protects the spinal cord inside the backbone. Because the lower vertebrae carry more weight, they are thicker and heavier than the vertebrae higher up.

MORE THINGS TO DO

Knocking down walls

Try building some walls out of different materials and test how strong they are. Make your walls from plastic or wooden building blocks, paper cups, yogurt cups, books, beverage cans or cardboard tubes. What is the best way of arranging each material to make the strongest wall?

To test the strength of your walls, slide a toy car down the slope and record how many pieces of the wall are knocked out by the force of the car hitting the wall. With a steeper slope, how many more pieces are knocked out of the wall?

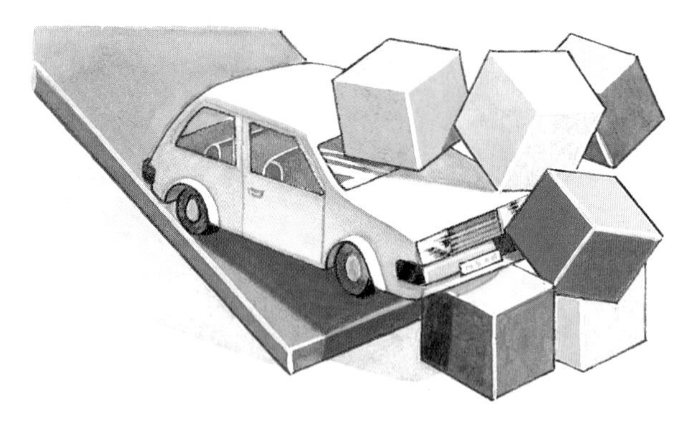

Another strength test would be to tie a string to a yogurt cup, half fill it with sand, and swing it against the walls. How far back do you have to hold the pot before you let go in order to knock down the whole wall?

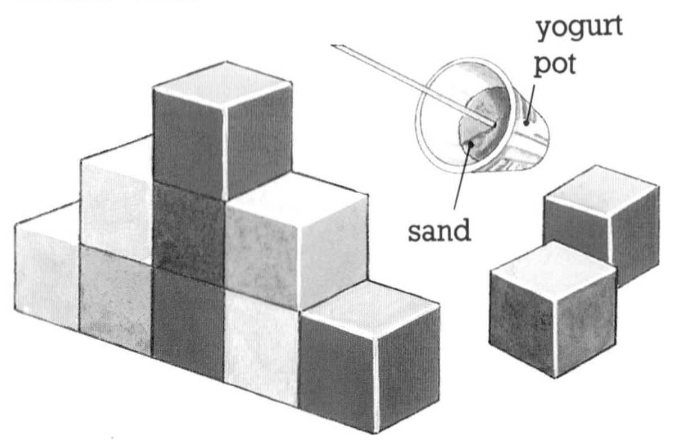

Eggshell strength

Place four halves of empty eggshells on a newspaper on a table with the dome shape of each shell pointing upward. Lay a piece of cardboard on top of the eggshells and arrange the shells so there is one at each corner of the cardboard. Carefully lay a heavy book on top of the cardboard. How many books can you pile on top of the first one before the eggshells break?

The pressure of the books follows the curve of the dome shape downwards. This spreads out the pressure and allows the shells to support a great weight.

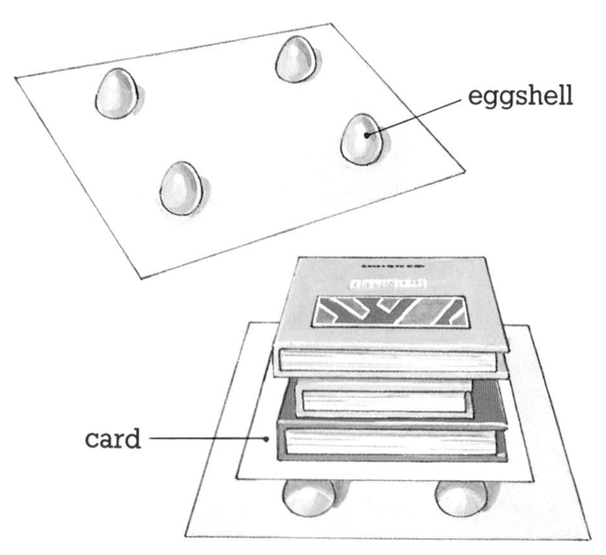

Wearing away stone

Put a piece of natural chalk (not blackboard chalk), limestone or mortar into a container and pour in enough vinegar to cover it. Can you see bubbles? Vinegar is a kind of acid that eats away at chalk and limestone. As it does this, it produces a gas that forms tiny bubbles. In a similar way, acid rain eats away at the stone on buildings, gradually destroying them. Can you remember what causes acid rain (see page 13)?

Make a floating bridge

Fill a large, flat bowl or the bathtub half full of water and float several clean, empty plastic food containers on the water. You need enough containers to bridge the gap between one side of the bowl or bathtub and the other. The containers should be close, but not touching. Anchor each one to the sides of the bowl or bathtub, with string and modeling clay. Place some strips of cardboard across the containers to finish the bridge. This type of floating bridge is called a pontoon bridge. How much weight will your pontoon bridge support without collapsing?

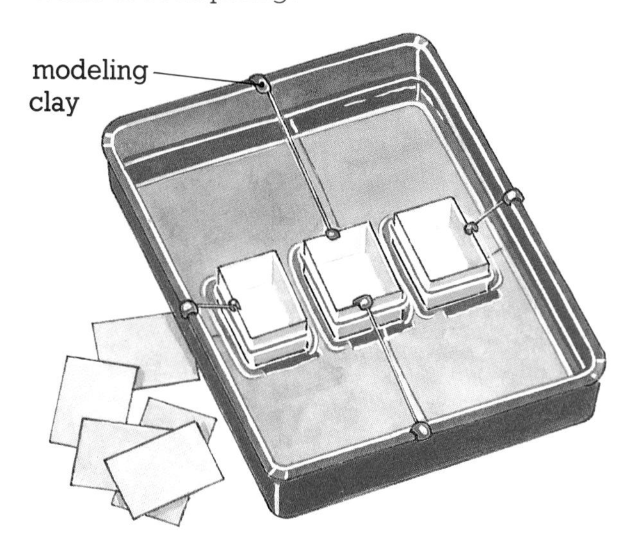

Collecting bricks

Make a collection of different types of brick. How many differences can you find? What shapes, sizes and colors are the bricks? How much do they weigh? Do they feel rough or smooth? Do any of the bricks have holes or dips in them? Which bricks do you like best? To make a record of your brick collection, you could place a piece of paper against each brick and rub gently over the surface with a crayon. Or you could press a thin layer of modeling clay onto the surfaces of the bricks and peel it off again.

Bird builders

See if you can find out about different kinds of birds' nests. Look up the information in books. You should never disturb a real bird's nest. What type of materials do the birds use and how do they join the materials together? Does the bird use the same nest year after year or build a new nest each year? How big is the nest?

Building buttresses

Use three pieces of cardboard to make a simple building with two walls and a flat roof. Put some weights on the roof. How much weight will the building take before the walls collapse? Now use wooden blocks as buttresses to prop up the walls on either side. How much weight will the roof take now? Which way do the walls collapse?

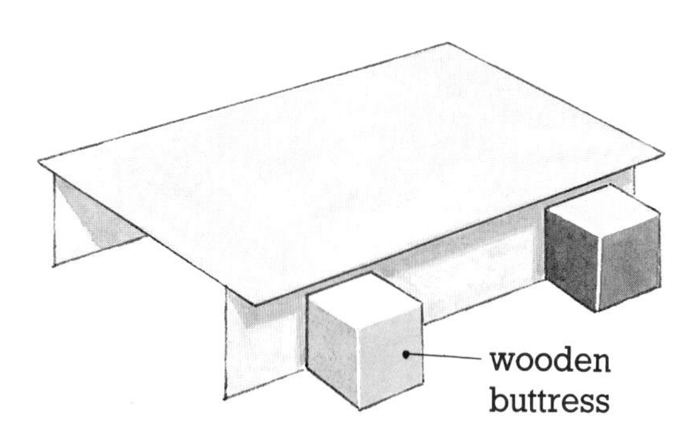

Swaying in the wind

Build a tall tower out of building blocks. To create a strong wind, blow a hairdrier (set on cold air) at the tower. Start some distance from the tower and gradually move closer. Does the tower sway in the wind? Change the angle of the hairdrier so the wind hits the tower from different directions. Does this make a difference? Can you think of a way of supporting your tower so it does not blow down easily in a strong wind? Would foundations help?

DID YOU KNOW ?

▲ The tallest building in the world (not the tallest structure) is the Sears Tower in Chicago. It is 1,452 feet high and has 110 stories, 103 elevators and 16,000 windows. About 17,000 people work in the building.

▲ There are 206 bones in the human skeleton. Just one of your hands contains 27 bones. For their weight, bones are stronger than steel or reinforced concrete.

▲ Weight for weight, the sticky silk of an orb spider's web is stronger than steel and more elastic than rubber. Each silken thread is a hundred times thinner than a human hair. Most spiders produce more than one type of silk; each type comes from a different silk gland in the spider's abdomen. The garden spider produces seven kinds of silk and each has a different use, from providing a tough casing for its eggs to supplying a strong safety line for the adult spider.

▲ The pyramids at Giza, Egypt, were built 4,500 years ago. In the Great Pyramid, there were over two million blocks of stone with an average weight of 2½ tons each. The pyramid was 485 feet high and took more than 20 years to build.

▲ Stonehenge, England, was begun in about 2,750 BC, nearly 200 years before the ancient Egyptians started work on the Great Pyramid. It probably took about 1,500,000 days to build.

▲ One of the stone blocks used by the Incas in a building weighed 138 tons. It would have taken 2,400 workers to put it into position.

▲ Some buildings are designed to sway more than others. The top of Moscow's concrete television tower, which is 1,760 feet high, sways as much as 19 feet from side to side in strong winds. The Sears Tower in Chicago has a more rigid frame that should never sway more than 3 feet. If a skyscraper sways too much, the people inside may feel "seasick."

▲ The Achilles tendon, at the back of your ankle, is no thicker than string. It bears all the tensile, or stretching, force from your calf muscles and allows you to walk and jump.

▲ The leaning tower of Pisa leans badly to one side because the foundations were not firm enough to support the weight of the tower. The tower has tilted 16½ feet from the perpendicular.

▲ The world's first iron bridge was built in 1779 at Ironbridge, England. It still stands today.

▲ The longest floating bridge is the Second Lake Washington Bridge in Seattle, Washington. The floating section of the bridge is 1.43 miles long.

▲ The longest bridge in the world is the Second Lake Pontchartrain Causeway in Louisiana. It is nearly 24 miles long and no land can be seen from the middle of the bridge. The bridge is made of concrete spans, each of which is 56 feet long.

▲ The widest bridge in the world is the Crawford Street Bridge in Rhode Island. It is 1,148 feet wide.

GLOSSARY

Acid rain
Rain that is much more acidic than normal because it has chemicals from vehicle exhausts, power stations and factories dissolved in it.

Arch
A curved structure in which the weight of the center is carried outward and downward to the ground.

Architect
A person who designs and oversees the building of a structure.

Beam bridge
A flat, rigid bridge that is supported at each end.

Buttress
A structure, usually made of stone or brick, built against a wall to prop it up and give extra support.

Cantilever
A beam that is free at one end and counterbalanced at the other end.

Compression
A pressing or squeezing force which tends to reduce or shorten things.

Flexible
Able to bend without breaking.

Force
A push or a pull that makes an object move or change its speed or direction.

Foundation
The strong base of a building which keeps it from sinking or leaning.

Materials
The substances from which things are made.

Molecule
The smallest particle of a substance that can exist by itself and still have the properties of that substance.

Pontoon bridge
A floating river or sea bridge which is supported by a series of boats or floats.

Reinforced concrete
Concrete which contains steel rods, bars, wires or mesh to give it extra strength.

Skeleton
The hard framework of bones, shell or tough skin that supports an animal's body from the inside or encloses it on the outside.

Span
The section of a bridge between two supports.

Suspension bridge
A bridge in which the roadway or deck hangs from steel cables supported by towers.

Tension
A force that pulls apart or stretches things and tends to lengthen them.

Answer to question on page 12: metal.

INDEX

Additional photographs:
Alan Cork 20 (b); The Environmental Picture Library 13 (b); Frank Lane Picture Agency 22 (t); N.R. Coulton/NHPA 9; Betty Rawlings Freelance 7 (t), Zefa 4, 8, 11, 17 (b), 22(b), 24.
Picture Researcher:
Ambreen Husain